ANIMAL GROUPS

A SLEUTH OF BEARS

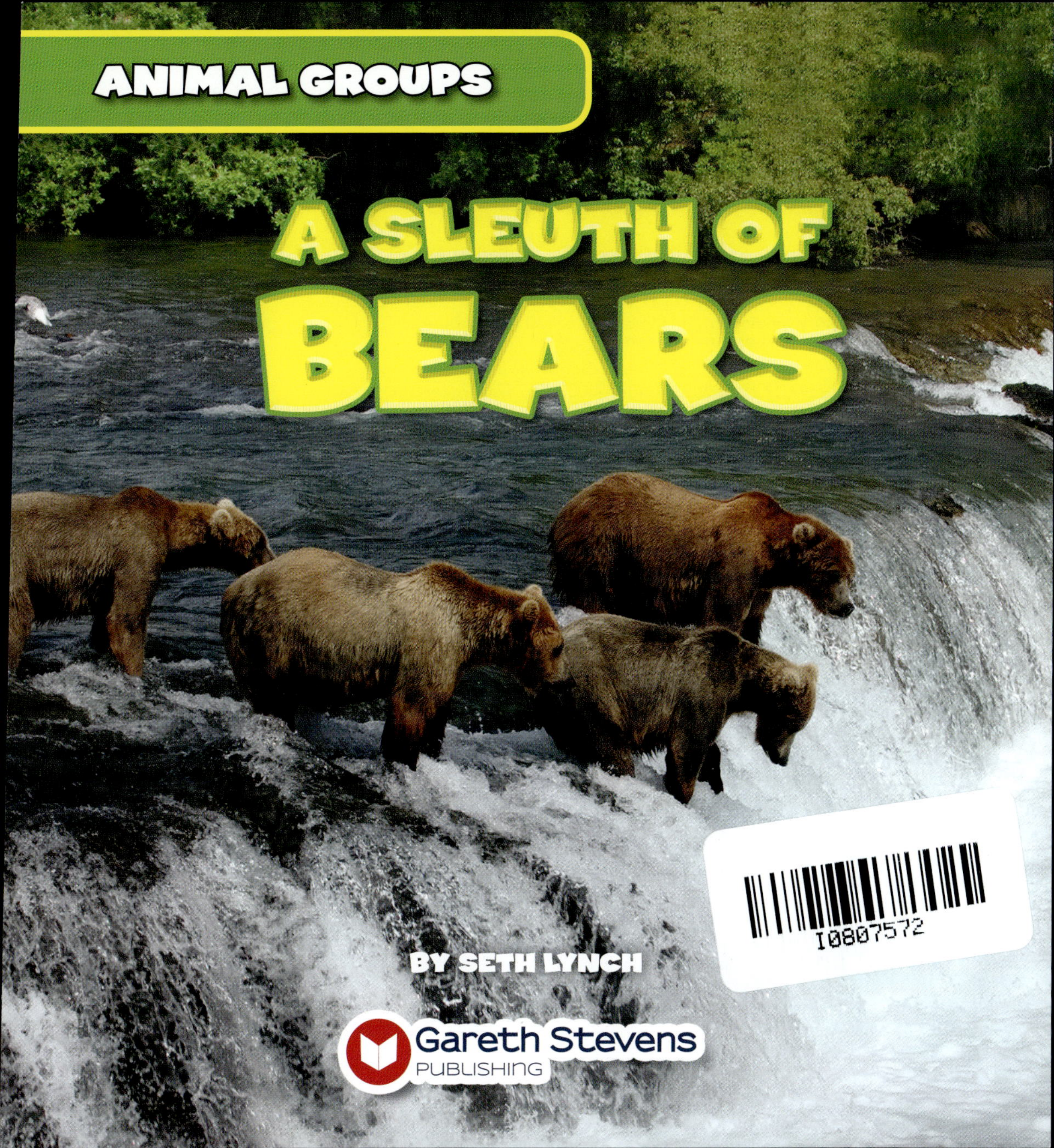

BY SETH LYNCH

Gareth Stevens PUBLISHING

Please visit our website, www.garethstevens.com. For a free color catalog of all our high-quality books, call toll free 1-800-542-2595 or fax 1-877-542-2596.

Library of Congress Cataloging-in-Publication Data

Names: Lynch, Seth, author.
Title: A sleuth of bears / Seth Lynch.
Description: New York : Gareth Stevens Publishing, [2022] | Series: Animal groups | Includes index. | Audience: Ages 6 | Audience: Grades K-1
Identifiers: LCCN 2020028910 (print) | LCCN 2020028911 (ebook) | ISBN 9781538263594 (library binding) | ISBN 9781538263570 (paperback) | ISBN 9781538263587 (set) | ISBN 9781538263600 (ebook)
Subjects: LCSH: Bears–Juvenile literature. | Bears–Behavior–Juvenile literature.
Classification: LCC QL737.C27 L946 2022 (print) | LCC QL737.C27 (ebook) | DDC 599.7815–dc23
LC record available at https://lccn.loc.gov/2020028910
LC ebook record available at https://lccn.loc.gov/2020028911

Published in 2022 by
Gareth Stevens Publishing
111 East 14th Street, Suite 349
New York, NY 10003

Designer: Tanya Dellaccio
Editor: Kristen Nelson

Photo credits: Cover davidkl/E+/Getty Images; p. 5 Yatra/Shutterstock.com; p. 7 BGSmith/Shutterstock.com; p. 9 Martin Mecnarowski/Shutterstock.com; p. 11 Sergey Uryadnikov/Shutterstock.com; p. 13 Dennis W Donohue/Shutterstock.com; p. 15 slowmotiongli/Shutterstock.com; p. 17 Sergey Uryadnikov/Shutterstock.com; p. 19 Volodymyr Burdiak/Shutterstock.com; p. 21 Yoreh/Shutterstock.com.

Printed in the United States of America

CPSIA compliance information: Batch #CSGS22: For further information contact Gareth Stevens, New York, New York at 1-800-542-2595.

CONTENTS

Boldface words appear in the glossary.

Many Bears

Bears live in North and South America, Europe, and Asia. There are many kinds. Some are large. Kodiak bears can weigh more than 1,000 pounds (454 kg)! Sun bears are small, growing to be only about 60 pounds (27 kg).

SUN BEAR

Though bears can look different, they share some **behaviors**. All bears are **omnivores**. But they don't eat the same things. Polar bears love to eat seals. Pandas mostly eat a plant called bamboo. American black bears really enjoy **insect** larvae and berries!

BLACK BEAR

On Their Own

A group of bears has a cool name: a sleuth! However, bears of all kinds usually live on their own. Sometimes good food, like a tasty berry patch, might draw several bears. They'll likely share it if there is enough food in the area.

BROWN BEARS

The Home Range

Bears have a home range they share with other bears. They might fight another of their kind for a **mate** or food, but they don't **defend** certain areas to keep others out. Male bears usually have a larger home range than females.

POLAR BEARS

Bears that share home ranges know each other. They may like some bears better than others, just like people. Even though bears don't live in groups, large adult males are still the top bears in an area.

GRIZZLY
BEAR

Cubs on the Way

Bears do come together in sleuths when it's time to mate. This only happens about once a year. Some bears only mate every few years! Baby bears, or cubs, may not be born for many months after mating.

Female bears wait until they are safely in a winter **den** to have their cubs. Bears may have up to four cubs at a time. Pandas usually only have one cub at a time. Mother bears feed and care for their cubs.

Growing Up

Cubs stay with their mother until the next time she's ready to have cubs. This can be over a year. As bears grow up, they might become friends with each other! Young bears may spend time in pairs or small groups to stay safe.

Bears in Danger

All kinds of bears are in trouble because of the activities of people. **Habitat** loss means there's less area to find food. Some bears are hunted too. Pandas are in the most danger. There are only about 1,600 left. It is up to us to protect bears!

PANDA

GLOSSARY

behavior: a way an animal acts

defend: to keep something safe

den: an animal's hidden home

habitat: the natural place where an animal or plant lives

insect: a small, often winged, animal with six legs and three main body parts

mate: one of two animals that come together to produce babies. Also, to come together to produce babies.

omnivore: an animal that eats both meat and plants

FOR MORE INFORMATION

BOOKS

Horning, Nicole. *Bear Cubs*. New York, NY: Cavendish Square Publishing, 2021.

Rose, Rachel. *Grizzly Bear*. Minneapolis, MN: Bearport Publishing Company, 2021.

WEBSITES

Brown Bear

kids.sandiegozoo.org/animals/brown-bear

Read about the great brown bear and check out pictures and a video.

For Teachers & Kids – Bears

www.nps.gov/subjects/bears/teacherskids.htm

Find links to videos and sound clips of bears in the wild from the National Park Service.

Publisher's note to educators and parents: Our editors have carefully reviewed these websites to ensure that they are suitable for students. Many websites change frequently, however, and we cannot guarantee that a site's future contents will continue to meet our high standards of quality and educational value. Be advised that students should be closely supervised whenever they access the internet.

INDEX